CONTENTS

INTRODUCTION

All the boxes, baskets, jars and tins shown in this book are intrinsically useful, but they've been taken a step beyond utility to become beautiful, decorative objects. Every home needs plenty of storage, and whether or not you relish tidying up and making everything shipshape, one way to make it easier is to have the right container to hold all your bits and pieces. If it looks good too, so much the better. It's Shaker philosophy to make nothing unless it is useful, but to make what is useful beautiful. While these designs may be too fanciful to suit the precepts of those restrained craftsmen, it's always true that using beautiful things adds to the pleasures of life, and if you've created them yourself they are twice as satisfying.

Specially made packaging makes the simplest gift a delight to receive, and shows real thought and care. Awkwardly shaped gifts, or a selection of small parcels, will all fit into a lidded box, with the added bonus that the box can be used again and again afterwards.

Several of these projects include a template, which you may need to enlarge to the size you need. If you have access to a photocopier it will do the job for you. Otherwise, draw a squared grid over the template and copy the pattern, square by square, on to a sheet of graph paper with a larger scale.

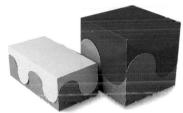

Use our ideas as a starting point for your own ingenuity, and adapt the techniques to suit your decorating style. Small-scale projects like these are an excellent way to sample a craft you haven't tried before. You can learn lots of new skills from the following pages, which are full of detailed instructions in simple steps that are easy to master.

WOODBURNT CANDLE BOX

Decorative woodburning belongs to the folk art of Scandinavia, where hot irons were used to scorch traditional patterns into wooden objects, particularly kitchenware, since the patterns had to withstand vigorous scrubbing. This design is borrowed from an early American candle box of Pennsylvania Dutch origin.

MATERIALS

soft pencil
tracing paper
scissors
plain pine box
chalk-backed transfer paper
masking tape
woodburning kit with chisel-ended and flat tools
paintbrushes
satin acrylic varnish
burnt sienna acrylic paint

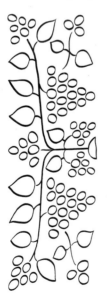

1 Trace, scale up and cut out the template to fit the plain pine box. Place the chalk-backed transfer paper between the tracing and the box, and secure with masking tape. Transfer the designs using a soft pencil. Remove the paper.

2 Set the woodburner at a medium heat. Using the chisel-ended tool, follow the pattern lines for the stems and outlines of the leaves. Keep the tool moving or lift it off, as it will burn more deeply if it is held in one place for too long.

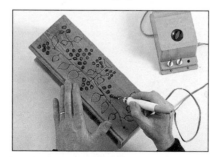

3 Use the chisel-ended tool with a prodding movement to fill the leaf outline with dots. Outline the fruit with the flat tool.

4 Finally, apply two coats of satin acrylic varnish tinted with a little burnt sienna acrylic paint. Follow with one coat of clear varnish.

STAMPED GIFT BOX

Boxes are the perfect packing for awkwardly shaped gifts. Use our simple pattern to transform a sheet of thin card (cardboard) into an attractive box, having stamped a repeating design across the card (cardboard) before folding it. Make it special by choosing a motif to suit the gift inside.

MATERIALS

pencil
tracing paper
metal ruler
card (cardboard) or paper for template
thin coloured card (cardboard)
craft knife
cutting mat
sheet of paper
rubber stamp
stamp inkpad
double-sided adhesive tape

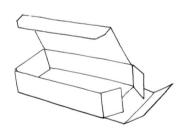

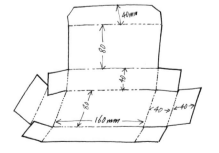

1 Using the template, scale up the measurements according to the size of your gift. Then draw the box shape on to the card (cardboard).

2 Cut out the box, using a craft knife and a metal ruler to keep the edges neat. Use the cutting mat to protect your work surface.

3 Using a straight-sided sheet of paper or a ruler as a guide, stamp your motif in diagonal rows across the card (cardboard). Extend the pattern to the edge by stamping partial motifs at the ends of each row, keeping the spacing constant. Allow the ink to dry.

4 On the wrong side of the card (cardboard) score along the fold lines with the back of the knife. Fold along the score lines, making sure the corners are square. Join the overlapping flaps on the inside of the box using double-sided adhesive tape. Fold in the end pieces and stick.

SUNFLOWER CANISTER

Transform plain metal storage canisters quickly by spraying them with aerosol paint and then splashing cheerful sunflowers all over them. This is an inexpensive alternative to buying painted storage jars in kitchen shops. A matching set of six or more will look great on display in the kitchen.

MATERIALS

*plain metal storage canister
matt blue spray paint
pencil
medium and fine paintbrushes
acrylic paints: yellow, orange, brown
and cream
acrylic sealant spray*

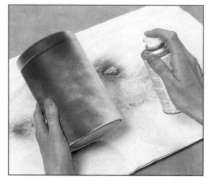

1 Wash the storage canister and lid to remove any grease. Dry thoroughly. Spray them with the matt blue paint, building the colour up in several thin coats. Allow each coat to dry before applying the next, to prevent the paint from running.

2 Using the acrylic paints, paint in the sunflowers. For each one, paint a yellow circle about 3 cm (1¼ in) in diameter and then space the petals evenly around the edge. Repeat the motif until the whole surface is covered. Allow to dry.

3 Add more colour to the yellow petals to give a feeling of depth. Paint the centres with circles of brown, and highlight with cream. Allow to dry, then spray the canister and lid with an acrylic sealant to protect the surface. The canister will withstand gentle cleaning, but do not put in the dishwasher.

TIP
Mark the position of all the flowers lightly with a pencil before you begin painting. This will ensure your design fits evenly around the canister.

SEASHELL BOX

Combine the contemporary look of corrugated cardboard with a dynamic shell arrangement. For the finishing touch, paint the box in pure white, and the result is a seashell box that resembles a meringue-topped cake.

MATERIALS

*selection of seashells
round corrugated cardboard box
with lid
glue gun (or all-purpose glue)
paintbrush
white acrylic gesso or paint*

1 Lay out all the seashells and sort them into different shapes and sizes. Arrange them on the lid of the round corrugated cardboard box.

2 Remove the top layer of shells and begin to glue the bottom layer on. Glue the outside shells first, and gradually move inwards.

TIP
It is best to use the larger shells on the bottom layer of your design, as they will provide a firm base for the smaller, more delicate shells.

3 Work with the shell shapes, building up the middle section. The glue gun provides an instant bond, so the shells will stick to the surface however you wish.

4 Paint the box and the lid white. If you are using acrylic gesso, two coats will give a good matt covering; ordinary acrylic paint will benefit from an extra coat.

PRESENTATION BOX

This lavishly decorated box is almost a gift in itself, and is a stylish way
to conceal a special surprise. Natural materials work well with fabric
flowers, and you can use paper ribbon both to cover the box and to tie
the chunky bow on top.

MATERIALS
....
scissors
paper ribbon
oval cardboard box with lid
PVA (white) glue
paintbrush
corrugated cardboard
pencil
selection of fabric flowers

1 Cut the paper ribbon long enough to fit around the rim of the box lid with a small overlap. Unfurl, and glue in place using a paintbrush. Fold and glue the excess ribbon under and above the rim.

2 Place the lid on the corrugated cardboard with the corrugations running straight from end to end. Draw around the edge and cut out the shape. Glue on to the lid to cover the surplus ribbon.

3 Cut two lengths of ribbon each measuring three times the width of the lid. Unfurl, and glue one end of each piece to the lid, tucking the ends under the rim. Tie the other ends together in a large bow. Trim.

4 Put the lid on the box. Measure the distance from the bottom of the box to the lower edge of the lid. Cut a strip of corrugated cardboard this width, and long enough to wrap around the box. Glue in place.

5 Cut the fabric flowers and leaves from their wire stalks. Glue on to the box, so that they give the effect of a sheaf of flowers lying underneath the bow.

ORIGAMI CARTON

Folded paper makes surprisingly strong boxes. For a different effect, you can use patterned paper instead of a painted design. Children will be enchanted by a party meal in the garden served in lots of these little containers.

MATERIALS

rectangular sheet of coloured paper
paintbrush
acrylic paints

1 Fold the paper in half twice in both directions to make 16 equal sections. Open out. Fold the two short sides in towards the centre.

2 Working from the short sides of the folded rectangle, fold in the four corners as far as the first crease.

3 Fold back the two centre strips over the four triangles.

4 Hold the sides in the centre and pull them up, then apart. Pinch the corners and the creases around the base to straighten them. Paint on your own design.

TIP

Choose fairly stiff paper that will hold sharp creases and, when cutting your rectangle, remember that the base of the finished box will be a quarter of the size of the sheet.

SEA HORSE STORAGE BOX

Delicate, swirling brush strokes create a watery background for this
delightful sea horse. The box can be used for storing stationery, pencils,
jewellery or even cosmetics.

MATERIALS
....

coping saw or fret saw
5 mm (¼ in) thick pine slat,
68 x 3 cm (27 x 1¼ in)
ruler
pencil
medium and fine-grade sandpaper
wood glue
masking tape
5 mm (¼ in) thick birch-faced
plywood sheet, 40 cm (16 in) square
medium and fine paintbrushes
white undercoat paint
acrylic paints: blue, white, green and
gold
tracing paper
stencil card (cardboard) or acetate
sheet
craft knife
cutting mat
stencil brush

1 Cut two 20 cm (8 in) and two 14 cm (5½ in) lengths of pine slat. Use the sandpaper to remove any rough edges. Glue together to form the sides of the box. Hold the frame together with masking tape while the wood glue is drying.

2 Cut two pieces of thick birch-faced plywood to fit inside the frame for the base and lid insert. Draw around the outside of the box and cut out to make the lid. Glue the base into the frame. Sand the edges of the lid insert and glue it to the lid.

3 Sand the box, paint it with white undercoat. When dry, sand the box again using fine-grade sandpaper. Paint the box with swirling brush strokes using two shades of blue acrylic paint.

4 Copy the template on to tracing paper, and transfer it to the stencil card (cardboard) or acetate sheet. Cut out carefully with a craft knife, using a cutting mat to protect your work surface.

5 Tape the stencil to the lid
and fill in the pattern using
a combination of blue and green
acrylic paints. Finish with a very
light smattering of gold. When the
paint is dry, lightly sand the whole
box with fine-grade sandpaper.

TOLEWARE CANISTER

Toleware, or painted tin, derives its distinctive character from bright colours and bold brush strokes, which contrast strongly with a dark background. The finish is permanent and, though not as durable as enamelled tin, is ideal for decorating storage containers.

MATERIALS

plain metal canister
coarse-grade steel wool
medium and fine paintbrushes
black matt emulsion (flat latex) paint
tracing paper
soft pencil
chalk-backed transfer paper
masking tape
acrylic paints: red, forest green, white, yellow-green, burnt sienna and raw sienna
satin acrylic varnish

1 Wash the metal canister to remove any grease or dirt. Dry thoroughly, and rub down with coarse-grade steel wool to provide a key for the paint. Apply two coats of black matt emulsion (flat latex) paint and leave to dry.

2 Design your own template on the tracing paper to fit the canister. Place the chalk-backed transfer paper between the tracing paper and the canister and secure with masking tape. Draw around the outlines with a soft pencil to transfer the design.

3 Thin the acrylic paint until the brush glides easily, and begin by filling in the round, red shapes. Practise on paper before painting the leaves forest green, beginning with a light touch and then applying more pressure to spread the brush for the thicker end of the stroke.

4 Don't overdo the highlights: just skim the surface, following the curves of the pattern. Use white on the red areas and yellow-green on the green areas. Leave to dry.

5 Tint some of the varnish with a small amount of both burnt and raw sienna. Apply two even coats, followed by one coat of clear acrylic varnish and leave to dry.

DECOUPAGE VALENTINE'S BOX

*A romantic keepsake for Valentine's Day, or any time of the year. Avoid the
usual whimsical "Victoriana" look by using strong colours and shapes.
Découpage is an easy technique that gives spectacular results.*

MATERIALS

*plain wooden box
medium paintbrushes
red emulsion (flat latex) paint
scissors
floral wrapping paper
pencil
tracing paper
card (cardboard) or paper for template
gold paper
PVA (white) glue
antique oak varnish
soft cloth
polyurethane gloss varnish*

TIP
Try out the whole of your
découpage arrangement before you
start to glue any shapes on to
the box.

1 Paint the box red inside and out.
Cut out flower images of various
sizes from the floral paper. Enlarge
the heart template and cut it out of
card (cardboard) or paper. Use it to
cut eight hearts out of the gold paper.

2 Arrange the hearts on the box,
two on the lid, two on each long
side and one on each end. Glue them
in place. Glue the flower images
around the hearts, pushing out
any air bubbles. Leave to dry.

3 Add a coat of antique oak
varnish and rub it off with a
cloth, to give an old, soft look. Then
finish with three or four coats of
polyurethane gloss varnish, leaving
each to dry before applying the next.

GILDED SUN AND MOON BOX

A gilded sun graces the lid of a plain wooden box with a touch of celestial mystery. This luxurious effect is easy to achieve using Dutch metal leaf. Delineate the area to be gilded with Japanese gold size to provide an adhesive backing, then apply the Dutch metal leaf as a transfer.

MATERIALS

medium and fine paintbrushes
wooden box with lid
acrylic gesso
ultramarine acrylic paint
sandpaper
gloss varnish
Japanese gold size
Dutch metal leaf transfer book
silver leaf transfer book

TIP
Too many coats of acrylic gesso may stop the lid from closing properly, so check that it still fits while you are painting the box.

1 Using the medium brush, paint the wooden box and lid, both inside and out, with three or more coats of acrylic gesso. Leave to dry between each coat.

2 Using the fine brush, give the box a coat of ultramarine acrylic paint. When dry, sand lightly for a distressed effect. Add a coat of varnish. Leave to dry.

3 Using the fine brush, paint a freehand sun motif on the lid with the Japanese gold size. When the surface is just tacky, place the Dutch metal leaf transfer on top and rub gently with a finger. Using the same technique, paint loose freehand moons around the side of the box. Paint the side of the lid with size, apply silver leaf transfer and leave to dry thoroughly.

STRING BOTTLE

*Liqueur bottles have such lovely shapes that it seems a shame to put them
into the recycling bin. This method of recycling enables you to go on
enjoying the bottles long after you have consumed their original contents.*

MATERIALS

*ball of thick string
glue gun (or all-purpose glue)
bottle
scissors*

1 Coil the string around to make a flat mat. Place a dot of glue in the centre of the bottle base, and spread it outwards, in spokes, towards the edge. Press the coiled string on to the base.

2 Glue around the lower edge to secure the base, then circle the bottle with the string, working your way up and gluing as you go. Keep the string taut, and make sure you get a good bonding on the bends.

3 When you reach the top of the bottle, cut the end of the string and apply plenty of glue to it so the finish is neat with no fraying. Wipe off any surplus glue that may have spread from beneath the string.

TIP
Don't use a bottle with a concave base. Check that the bottle stands firmly on its string base before you start winding string up the sides.

GOLD-RELIEF JARS

Use the brilliant colours of glass paints and gold outliner to create a rich and luminous effect that will look stunning by candlelight. The relief outliner produces a raised design with an almost Indian feel.

MATERIALS

jars
methylated spirits (rubbing alcohol)
medium paintbrush
solvent-based glass paints
gold glass-painting outliner
paper towels

1 Wash the jars in a solution of detergent to remove grease and sticky labels. Use methylated spirits (rubbing alcohol) to remove any stubborn bits. Dry thoroughly. Brush the solvent-based glass paints generously on to the jars, but do not let heavy drips build up.

2 Leave to dry for about 24 hours, then apply the designs with the gold glass-painting outliner. Start with the border at the top and bottom. If you make a mistake, you can wipe it away quickly with paper towels. The outliner takes a very long time to dry completely; though it is touch-dry within 24 hours, allow up to 72 hours to be really sure. Now, using the glass paints, fill in the motifs in the middle.

PLANT POT IN RELIEF

*Customize a white plant pot with a relief design by picking out the raised
details in ceramic paints. This is easy and fun to do, and you can choose
shades that match your own colour scheme perfectly.*

MATERIALS

*china pot with relief design
fine paintbrush
ceramic paints
polyurethane gloss varnish*

1 Wash the china pot in a solution
of detergent to remove any
grease or grime. Dry thoroughly.
Beginning at the top of the pot,
apply the ceramic paint carefully
with a fine paintbrush.

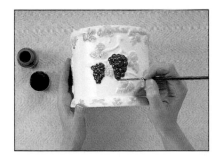

2 Following the raised details on
the china pot, paint in the main
motifs. Wipe away any mistakes
immediately before the paint dries.

3 When the decoration is complete,
leave the paint to dry completely.
Then cover with a protective coat of
polyurethane gloss varnish.

CURVY BOX

This unusual box is almost as pretty in two halves as it is when assembled.
For maximum impact, choose two contrasting or complementary colours for the
two halves so that the pattern of the curves stands out.

MATERIALS

pencil
tracing paper
card (cardboard) or paper for template
metal ruler
thin card (cardboard) in two
different colours
craft knife
cutting mat
glue stick

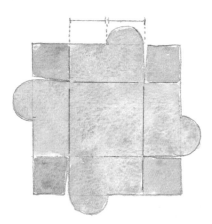

1 Scale up the template and transfer the pattern to each piece of card (cardboard). Cut out with a craft knife, using the cutting mat to protect your work surface.

2 Score the fold lines with the back of the knife. Both halves of the box are made in the same way: fold each tab in behind the adjacent semicircle to form the sides, and glue.

TIP

If you want to make a rectangular box, simply double the width of the box base and repeat the semi-circular pattern on the long sides.

3 To assemble, interlock the two halves, making sure that each semicircle overlaps on the outside of the box.

PUNCHED TIN CANDLE HOLDER

Tin-punching is a traditional craft, typical of the folk art of America, that is now fashionable again, and it's easy to do. The effect is graphic yet delicately detailed, and the metal reflects the warm glow of candlelight.

MATERIALS

pencil
tracing paper
card (cardboard) or paper for template
aluminium or tin sheet
marker pen
gloves
tin snips or old scissors
magazine or newspaper
large, strong needle
tack hammer
metal ruler
epoxy resin glue
wire brush
candle

TIP
Remember never to leave burning candles unattended.

1 Enlarge the template, transfer it on to card (cardboard) or paper and cut it out. Draw around it on the aluminium or tin sheet with a marker pen. Wearing gloves, cut it out with tin snips or old scissors.

2 Lay the shaped metal on a magazine or newspaper to protect your work surface. Using a large, strong needle and a tack hammer, punch the pattern gently into the metal sheet.

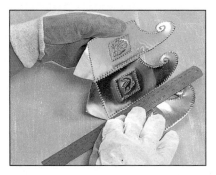

3 Fold the two outer metal panels inwards along the dotted lines, using a metal ruler to crease the sheet cleanly. Do the same with the triangular flaps at the bottom.

4 Secure the triangular shape and glue it in place. Scratch the surface all over with the wire brush. Put the candle in the bottom of the container.

FABRIC-LINED BASKET

This pretty basket is painted and lined with Provençal-style floral cotton to provide storage for the pots and tubes that gather in the bathroom.

MATERIALS

small round basket
fine-grade sandpaper
paintbrushes
acrylic paint in white and
colour to match fabric
matt (flat) varnish
tape measure
floral-printed cotton fabric
scissors
sewing machine
matching sewing thread
iron
1 m (3 ft 3 in) broderie anglaise
(eyelet lace), 6 cm (2¼ in) wide
lining fabric
sewing needle
2 m (6 ft) narrow woven floral ribbon
bodkin

1 Sand off any rough edges on the basket, and apply a thin coat of white acrylic paint. Leave to dry. Paint the rim and base of the basket in your chosen colour and leave to dry. Then apply a coat of matt (flat) varnish. Leave to dry.

2 Measure the circumference and radius and add 5 cm (2 in) to each figure. Cut the floral fabric, in a rectangular shape, to this size. Join the short edges, then press under 1 cm (½ in) along one side and topstitch to the broderie anglaise.

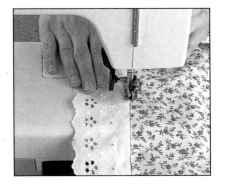

3 Cut the lining as long, but three times as wide, as the fabric. Join the short edges. Press under 1 cm (½ in) along one side and, wrong sides together, topstitch to the lace-trimmed edge of the fabric, 1 cm (½ in) below the previous stitching.

4 Using running stitch, sew along the lower edge of the lining. Gather and secure tightly with the excess fabric on the wrong side. Press under 1 cm (½ in) along the remaining side of the floral fabric and sew to the inside rim of the basket. Cut the ribbon in half, and use the bodkin to thread both pieces through the channel where the lining and the floral fabric meet. Knot the ends and draw to close.

Tip
When stitching the fabric to the broderie anglaise join it halfway down the lace. Also make sure the ends of the lace are neatly joined.

CHECKERED SEWING BOX

*A useful, pretty papier-mâché box with compartments for sewing
equipment. The surface pattern is created in the wet paint using a simple
comb, which you can cut from a spare piece of card (cardboard).*

MATERIALS

*thick card (cardboard) cut to the
following sizes:
base and lid 15 x 30 cm (6 x 12 in)
2 long sides 3 x 30 cm (1¼ x 12 in)
2 short sides and 3 partitions
3 x 14.5 cm (1¼ x 5¾ in)
4 lid tabs 1 x 4 cm (½ x 1½ in)
ruler
cutting mat
craft knife
gummed tape
newspaper
wallpaper paste
paintbrush
acrylic gesso or matt emulsion
(flat latex) paint
acrylic paint
card (cardboard) comb
varnish*

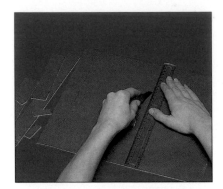

1 Cut the box pieces out of the
card (cardboard) using a craft
knife and ruler. Use the cutting mat
to protect your work surface.

2 Tape the box together. Fold the
tabs in half and position in the
corners of the lid, so that they will
fit inside the base of the box.

3 Soak some narrow strips of
newspaper in wallpaper paste
and cover the box and lid with two
layers of papier-mâché. Leave to dry.

4 Apply two coats of acrylic gesso
or matt emulsion (flat latex)
paint and leave to dry.

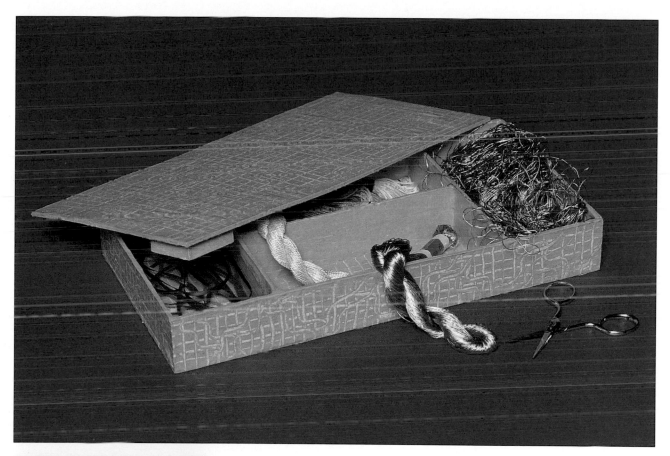

5 Apply the acrylic paint to one
section at a time. Mark the paint
into checks with the card (cardboard)
comb while it is still wet. Leave to
dry, then apply two coats of varnish.

Mexican Kitchen Box

This handy container for kitchen implements is made of salt dough, which is very durable as long as it doesn't get damp. It needs long, slow baking to make it rock hard and ready to decorate with vibrant colours.

MATERIALS

FOR THE SALT DOUGH:
10 cups plain (all-purpose) flour
5 cups salt
5 cups water
mixing bowl
rolling pin
baking parchment
pencil
tracing paper
card (cardboard) or paper for templates
scissors
small, sharp knife
fruit corer
baking tray
non-stick tart tray (muffin pan)
dressmaker's pin
fine-grade sandpaper
books or weights
string
paintbrushes
acrylic gesso or matt emulsion
(flat latex) paint
acrylic or craft paints
satin varnish
coloured raffia
strong, clear glue

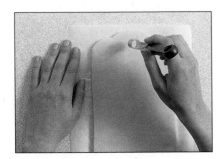

1 Mix together the flour, salt and half the water. Knead together, gradually adding more water until the dough is smooth and firm. Knead for a further 10 minutes. Roll the dough out on baking parchment to a thickness of 1 cm (½ in). Scale the templates to the required size and cut out of card (cardboard) or paper. Place on the dough and cut out the back, two sides and front. Cut out a square for the base. Use a fruit corer to punch a hole in the centre top of the back panel. Transfer all the pieces to a baking tray and bake at 120°C/250°F/Gas ½ for 1 hour.

2 To make a sombrero decoration, roll two balls of dough, each about 3 cm (1¼ in) in diameter. Flatten one ball for the brim and press into the cup of a tart tray (muffin pan). Flatten the base of the second ball and squeeze the top into a blunt point. Fix to the brim. Make two more and bake for 2¼ hours.

3 Roll three balls of dough 1.5 cm (⅝ in) and one ball 1 cm (½ in) in diameter and mould into flat egg shapes. Join to form a cactus. Use three tiny balls for the flowers and indent with the head of a pin. Prick the cactus to suggest spines. Make six more. Shape a 2 cm (¾ in) diameter ball into a chilli pepper. Roll a thin sausage to form a stalk, moisten one end and fix to the chilli. Make eight more chillies. Bake the cacti and chillies for 25 minutes until slightly hardened.

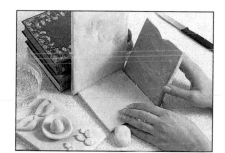

4 Moisten the surfaces of the motifs and attach to the box pieces with raw dough. Return to the oven, with the box base, for 18 hours. Remove, cool and sand the edges. Moisten the edges of the base and smear them with dough. Moisten the inner edges of the back, sides and front and assemble the box. Support the panels with books or weights as you position them. Tie string around the box to hold all the panels in place and return to the oven for 3 hours. Remove and sand along the joins. Undercoat the box with acrylic gesso or matt emulsion (flat latex) paint, then paint with acrylic or craft paints. Leave to dry. Apply five coats of satin varnish. Leave to dry. Tie a piece of coloured raffia around each sombrero and secure with strong, clear glue.

PERSONALIZED HAT BOX

*Vividly patterned oval pasteboard boxes were the early nineteenth-century
equivalent of modern hand luggage. A lady's hats, wigs, collars and ruffs were
safely stowed in these rigid containers.*

MATERIALS

drawing pins (thumb tacks)
tape measure
corrugated cardboard
strong thread
marker pen
cutting mat
craft knife
thin card (cardboard)
paperclip
PVA (white) glue
sticky tape
scissors
wallpaper
braid (trim)
wrapping paper

TIP

Use a paperclip to hold the side of
the box in position while you fit it
to the base.

1 To draw an oval base, place
two drawing pins (thumb tacks)
10 cm (4 in) apart on the corrugated
cardboard. Loop the thread around
both pins, place a marker pen within
the loop and, keeping it taut, draw
around the pins.

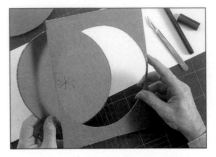

2 Cut out the base with a craft
knife. Draw around the base
section on another sheet of corrugated
cardboard. Cut this second piece out
around the outside of the line, making
a slightly larger oval for the lid of
the box.

3 To make the box, cut the
thin card (cardboard) as long as
the circumference of the base with an
overlap, and as wide as the required
depth. Glue the sides together, then
fix the base with sticky tape.

4 Cut a strip of wallpaper for
the side of the box, allowing it
to overlap 1.5 cm (⅝ in) at the top,
and at least 2.5 cm (1 in) at the base.
Snip the overlap, glue and turn in at
the top and base.

5 Place the lid on the wallpaper and, allowing a 2.5 cm (1 in) overlap, cut around it. Make the lid as you made the box, but use a strip of thin card (cardboard) 4 cm (1½ in) deep. Glue the wallpaper to the lid, snip the overlap and glue it to the side. Cut an oval of wallpaper to fit the base. Glue it to the base to hide the overlaps.

6 Glue the trimming to the side of the lid, or, if you prefer, use another strip of matching wallpaper. Then line the box with wrapping paper by measuring and sticking it in the same way as the outer paper.

LEO LETTER HOLDER

This jolly letter rack, emblazoned with its confident lion, is just the thing to brighten up your desk. Based on the astrological sign, it makes a thoughtful birthday gift for a Leo friend.

MATERIALS

pencil
6 mm (¼ in) birch plywood sheet cut to the following sizes:
base 21.5 x 7.3 cm (8½ x 2⅞ in)
sides 13 x 7.3 cm (5 x 2⅞ in)
front 23 x 10 cm (9 x 4 in)
back 23 x 19 cm (9 x 7½ in)
ruler
tracing paper
card (cardboard) or paper for template
coping saw or fret saw
sandpaper
wood glue
masking tape
four 1.5 cm (⅝ in) wooden balls
paintbrushes
stencil brush
white undercoat paint
acrylic paints: deep cobalt, deep yellow, cadmium red, gold, raw umber and black
matt (flat) varnish

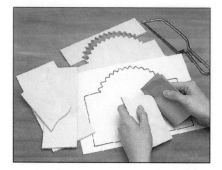

1 Mark out the back, front, base and two sides of the rack on the plywood to the sizes listed. Scale up the templates for the back and sides to the required size. Cut out all the pieces and sand the edges.

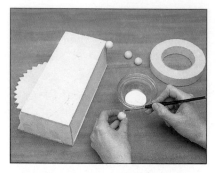

2 Glue the pieces together, holding in place with tape until the glue has hardened completely. Remove the tape and sand all the edges and corners. Glue the wooden balls to the corners of the base.

3 Paint the letter rack with white undercoat paint and sand down lightly when dry. Using the stencil brush, stipple the rack all over with deep cobalt acrylic paint.

4 Complete the design using the rest of the acrylic paints, using the main picture as a guide. Leave to dry. Seal and protect with a coat of matt (flat) varnish.

DECOUPAGE CAKE TIN

*This metal cake tin has been given a new lease of life with some white
paint and scrummy tartlet motifs – you could, of course, use your own
favourite cakes for the design instead.*

MATERIALS

*metal cake tin
paintbrush
white acrylic or matt emulsion
(flat latex) paint
card (cardboard) or paper for template
pencil
scissors
coloured cartridge (heavy) paper
PVA (white) glue
gloss varnish*

1 Wash the metal cake tin to
remove any grease and dry.
Prime the outside with white paint,
leave to dry, then apply a second
coat to give a dense base colour.

2 Make a template of your cake
design on the card (cardboard)
or paper and draw around it on the
coloured cartridge (heavy) paper.
Make enough cakes to cover the tin.

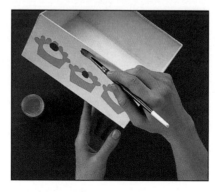

3 Arrange the cut-outs on the sides
and lid of the tin. Stick them in
position with PVA (white) glue.

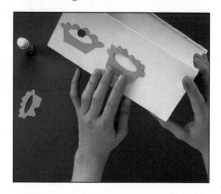

4 Seal the tin with two coats of
gloss varnish. Allow to dry
thoroughly before you use the tin.

TIP

For a different finish, paint your
designs using coloured acrylic or
matt emulsion (flat latex) paint.

CRAB BATHROOM BOX

This attractive little wooden box is very easy to construct, but the delicate painting and the raised crab design make it unusual and eye-catching. The lid is decorated with a wavy pattern inspired by the crab's watery home. It makes an ideal receptacle for your bathroom clutter.

MATERIALS

8 mm (⅜ in) thick pine slats,
40 x 3 cm (16 x 1¼ in)
coping saw or fret saw
4 mm (⅙ in) thick birch plywood

base and lid insert, 8 x 10 cm
(3¼ x 4 in)
lid, 11.5 x 10 cm (4½ x 4 in)
crab motif, 10 cm (4 in) square
ruler
pencil
tracing paper
sandpaper
wood glue
masking tape
paintbrushes
white undercoat paint
acrylic paints: blue, gold and red
matt (flat) varnish

1 Cut four 10 cm (4 in) lengths of pine slat and the base, lid and lid insert from the plywood. Enlarge the crab template and transfer it to the plywood. Carefully cut it out of the plywood. Use the sandpaper to remove rough edges.

2 Assemble the sides of the box and stick with wood glue. Hold the sides in place with masking tape until the glue is completely dry. Glue in the base. Glue the lid insert centrally on to the lid. Sand down any rough edges.

3 Paint the box and crab with a coat of white undercoat. Sand lightly when dry. Paint the box and lid blue, watered down and applied with a wavy brush stroke. Paint on the border pattern and stars. Paint the crab in red and pick out details in blue and gold. Finish off with a coat of matt (flat) varnish. When dry, glue the crab firmly on to the lid of the box.

PRIVATE LETTER BOX

This ornate box can be used to store your secret letters, or to keep other small documents safe. Use brightly coloured wrapping paper to decorate it, saving oddments (scraps) for extra details.

MATERIALS

pencil
tracing paper
thin card (cardboard)
pencil
craft knife
metal ruler
cutting mat
strong clear glue
masking tape
paintbrush
poster paints
wrapping paper
diluted PVA (white) glue
narrow ribbon
button

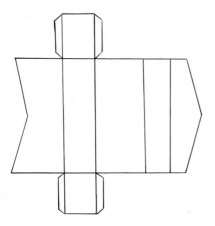

1 Scale up the template and transfer to the thin card (cardboard). Cut out using a craft knife and a metal ruler. Use the knife to score along the fold lines. Fold in the tabs and glue the sides.

2 When the glue is dry, strengthen the edges and sides of the box with strips of masking tape. Position them carefully and trim the corners to make a neat border. Paint the edging in a strong colour.

3 Cut out some motifs from the wrapping paper and arrange them on the box. Stick them down with diluted PVA (white) glue, and brush an extra coat all over the pieces to varnish them. Allow to dry. Fix a loop of narrow ribbon under the flap of the lid and a button on the front of the box to fasten.

TINY TRINKET BOX

This little triangular tartan fabric box, luxuriously lined with silk, will make a delightful gift either on its own or containing a tiny surprise. Cut the fabric for the rectangular sides on the bias to add interest to the design.

MATERIALS

pencil
tracing paper
thin card (cardboard) for templates
ruler
fusible woven heavyweight iron-on interlining, 50 cm (20 in) square
scissors
tartan fabric, 25 cm (10 in) square
silk fabric for lining,
25 cm (10 in) square
brushed cotton for interlining,
50 cm (20 in) square
iron
matching sewing thread
sewing needle

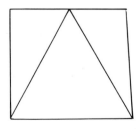

TIP

Use just the tip of the iron to press the seams on to the interlining.

1 Scale up the triangular and rectangular templates separately and cut them out of the thin card (cardboard). Using the templates, draw four triangles and six rectangles on the fusible woven heavyweight iron-on interlining. Cut them out.

2 Cut two triangles and three rectangles from the tartan fabric, leaving a 6 mm (¼ in) seam allowance all around. Repeat with the silk lining fabric. Cut four triangles and six rectangles from the brushed cotton, without any seam allowances.

3 Lay the tartan pieces wrong side up. Cover each with a brushed cotton piece, then an iron-on interlining piece, sticky side up. Fold the seam in and press with the iron.

4 Repeat step 3 with the silk pieces to form the lining. Centre on the tartan pieces, wrong sides together, and press until they have fused. Leave to cool. Whip stitch the edges.

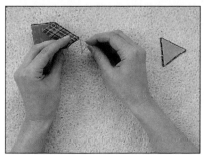

5 Whip stitch the three sides of the box to the base, right sides together. Pull up the sides of the box and slip stitch together. Slip stitch along the hinge to attach the lid.

CUTLERY BOX

This project fuses the clarity of high-tech design with the originality of surrealist sculpture and provides an ideal place to keep your cutlery. Make separate boxes for knives, forks and spoons and say goodbye to rummaging in the kitchen drawer.

MATERIALS

small silver-plated knife, fork and spoon, polished
3 metal boxes with lids
marker pen
metal file
coarse-grade sandpaper
metal bonding compound
craft knife

TIP

When you roughen the surfaces of the metal box lids to give a key for the metal bonding compound, be careful to do this accurately so that the roughened areas are completely hidden by the cutlery handles.

1 Bend the small silver-plated knife to form a right angle halfway along the handle. If it does not bend easily, use the edge of your work surface as a support.

2 Place the knife on one metal box lid and mark its position. Roughen the lid with a metal file and the contact point on the knife with coarse-grade sandpaper.

3 Mix the metal bonding compound, following the manufacturer's instructions. Apply to the rough area on the lid. The knife is fixed to the lid only at this point, so the bond needs to be strong.

4 Press the knife handle into position on the bonding compound. Use a fine instrument, such as a craft knife, to remove any excess. Repeat steps 1 to 4 for the small silver-plated fork and spoon.

FRENCH BREAD BIN

A rustic bread bin will bring a touch of French country style to your home. A carpenter could make the bin for you or, if woodworking is a hobby, make it at home. The decorative technique is known as ferning.

MATERIALS

reclaimed pine floorboards,
cut to the following sizes:
back: 28 x 76 cm (11 x 30 in)
front: 13 x 60 cm (5 x 24 in)
sides: 19 x 60 cm (7½ x 24 in)
pencil
tracing paper
jig-saw or coping saw
wood glue
panel pins
hammer
paintbrush
shellac
newspaper
masking tape
spray adhesive
selection of artificial ferns in
plastic or silk
spray paint in black, dark green or
dark blue
fine-grade sandpaper
matt (flat) varnish

1 Mitre the edges of the front and side pieces of reclaimed pine floorboards. Enlarge and trace the pattern for the back detail and cut it out using a jig-saw or coping saw. Join the pieces with wood glue. Use panel pins to secure. Cut the base to size and secure. Apply two coats of shellac to seal and colour the wood.

2 Working on one side of the bin at a time, mask off the surrounding area with newspaper and masking tape. Apply spray adhesive to one side of the pieces of artificial fern and arrange them on the surface. Make sure all the leaves are stuck closely to the wood to give a clear outline.

3 Spray on the colour, using light even sprays and building it up gradually. When the paint is dry, remove the ferns.

4 Work on the sides and the inside back panel in the same way until the whole bin is covered. Leave to dry.

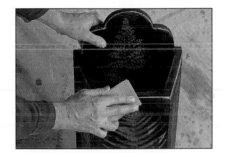

5 Sand the edges of the bin to
simulate a time-worn look.

6 Finally, apply two coats of matt
(flat) varnish to protect and seal
the fernwork.

PAINTED CHEST

You can use this pattern, based on an early American dowry chest, to decorate either an old or new blanket chest. The combing and spotting decorations in the varnish have to be done quickly, so work on one panel at a time.

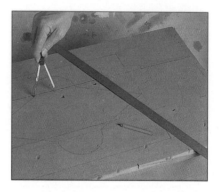

1 If you are starting with bare wood, first apply a coat of shellac to seal the surface, then paint the whole chest with dusky blue matt emulsion (flat latex) paint.

2 Trace and scale up the template and use it as a guide to position the panels on the chest. Draw them accurately, using a ruler, a pair of compasses and a pencil.

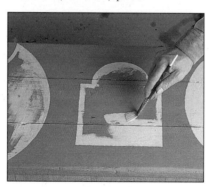

3 Fill in all the panels with cream matt emulsion (flat latex) paint. Leave to dry.

4 Apply a thick coat of antique pine varnish to one panel only. Do not allow to dry.

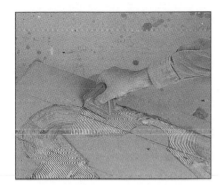

5 Quickly comb the varnish in a
regular pattern, following the
shape of the panel in one smooth
movement, then wipe the comb to
prevent any build-up of varnish.
Repeat step 4 and 5 on other panels.

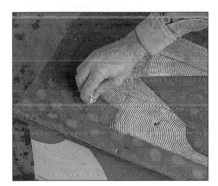

6 When all the panels are dry,
apply another coat of varnish
to the whole chest. Before the
varnish dries, take a just-damp
cloth, screw it into a ball and use
it to dab off spots of the varnish.

MEDITERRANEAN CRATES

Wooden fruit and vegetable crates are much too good to be thrown away once emptied, so rescue them and dress them up with vibrant colour and ribbon to make a great set of boxes for candles or table-linen (linens).

MATERIALS

3 wooden fruit or vegetable crates
sandpaper
pliers
wire cutters
staple gun (optional)
powder paint: red, blue and green
paintbrushes
1 m (3 ft 3 in) checked ribbon
scissors

TIP
Visit your local greengrocer or vegetable market and pick the best crates available. The ones shown here work especially well, as they have a solid base that can be separated and used as a lid.

1 Rub off any rough edges on the wooden crates with sandpaper. Detach the base from one crate to use as a lid. Remove and replace any staples if necessary.

2 Mix the powder paints following the manufacturer's instructions. Paint one of the crates red inside, blue outside and green along the top edges.

3 Paint the other crate green inside, blue outside and red along the top edges.

4 Paint the lid blue. Bind the six centre joints with crosses of checked ribbon, and tie underneath.

STAR CUPBOARD

While the style of this little cupboard is individual, it doesn't scream for attention, but has a comfortable, lived-in look. This stamping technique is so simple that you might want to transform other pieces of furniture in the same way.

MATERIALS

wooden wall cupboard
paintbrushes
matt emulsion (flat latex) paint:
olive-green, off-white and vermilion
pencil
tracing paper
card (cardboard) or paper for template
craft knife
cutting mat
medium-density sponge
marker pen
scissors
plate
PVA (white) glue
antique pine matt (flat) varnish
dry cloth

1 Apply two coats of olive-green matt emulsion (flat latex) paint to the cupboard. Scale up the star template and cut out of thin card (cardboard) or paper. Place the template on the medium-density sponge, draw around it and cut it out. Spread some off-white matt emulsion (flat latex) paint on to a plate. Dip the star sponge shape into the paint and print stars all over the cupboard, placing them quite close together. Leave to dry.

2 Mix two parts vermilion matt emulsion (flat latex) paint with one part PVA (white) glue. Check the stars are dry, and coat the cupboard with a liberal amount of this colour, daubed on with a paintbrush.

3 Finish the cupboard with a coat of antique pine matt (flat) varnish but, before it dries, use a dry cloth to rub some of it off each star. This layering of colour gives the surface its rich patina.

BATHROOM BUCKETS

Add an element of seaside fun to your bathroom by hanging up this row of jolly buckets. These were bought from a toyshop, but you could take a trip to the seaside for a great selection of buckets in all shapes and sizes.

MATERIALS

3 enamelled metal buckets
length of driftwood
pencil
drill with wood and masonry bits
wire
pliers
wire cutters
masking tape
wallplugs and screws
screwdriver

1 Line the enamelled metal buckets up at equal distances along the length of driftwood. Make two pencil marks for each bucket, one at each end of the handle where it dips.

2 Using the wood bit, drill holes through the six marks on the wood to hang the buckets up.

3 Wind wire around each handle and poke it through the holes in the wood. Twist the ends together at the back to secure and trim the ends. Drill a hole at each end of the wood.

4 Hold the wood on the wall and mark the positions for the fixings. Place masking tape over the tiles to prevent them from cracking, then, using the masonry bit, drill the holes and fix the wood to the wall.

TIP
If you can't find the ideal bit of driftwood on a beachcombing outing, use a length cut from an old plank to retain the seaside character of this idea.

30 CREATIVE CONTAINERS

Inspirational ideas for making and decorating containers

ULTIMATE
EDITIONS

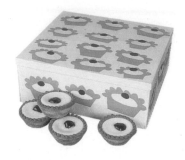

First published in 1997 by Ultimate Editions

© 1997 Anness Publishing Limited

Ultimate Editions is an imprint of
Anness Publishing Limited
Hermes House
88–89 Blackfriars Road
London SE1 8HA

ISBN 1 86035 247 2

Distributed in Canada by Book Express, an imprint of
Raincoast Books Distribution Limited

Publisher: Joanna Lorenz
Project Editor: Fiona Eaton
Designer: Lilian Lindblom
Illustrations: Anna Koska

Printed in China